NESSRINE SOLI
KHAOULA HIDOURI
BECHIR CHAOUACHI

Feasibility study of hydrocarbons as environmentally friendly refrigerants

NESSRINE SOLI
KHAOULA HIDOURI
BECHIR CHAOUACHI

Feasibility study of hydrocarbons as environmentally friendly refrigerants

Towards sustainable refrigeration

ScienciaScripts

Imprint

Any brand names and product names mentioned in this book are subject to trademark, brand or patent protection and are trademarks or registered trademarks of their respective holders. The use of brand names, product names, common names, trade names, product descriptions etc. even without a particular marking in this work is in no way to be construed to mean that such names may be regarded as unrestricted in respect of trademark and brand protection legislation and could thus be used by anyone.

Cover image: www.ingimage.com

This book is a translation from the original published under ISBN 978-3-8416-3230-2.

Publisher:
Sciencia Scripts
is a trademark of
Dodo Books Indian Ocean Ltd. and OmniScriptum S.R.L publishing group

120 High Road, East Finchley, London, N2 9ED, United Kingdom
Str. Armeneasca 28/1, office 1, Chisinau MD-2012, Republic of Moldova, Europe
Managing Directors: Ieva Konstantinova, Victoria Ursu
info@omniscriptum.com

Printed at: see last page
ISBN: 978-620-3-48953-8

Contents

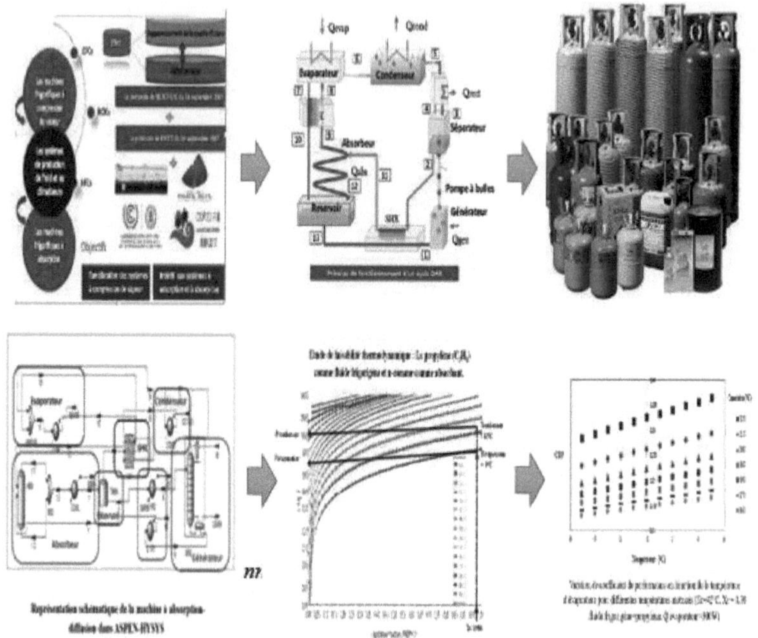

Représentation schématique de la machine à absorption-diffusion dans ASPEN-HYSYS

Variation du coefficient de performance en fonction de la température d'évaporation pour différentes températures motrices (Tc=42°C, Xr = 0,30 fluide frigo : gaz-propylène Qévaporateur=100W)

Acknowledgements

The present work was carried out at the Research Laboratory: Energy, Water, Environment and Processes (LEEEP) LR18ES35, at the National Engineering School of Gabes (ENIG) linked to the University of Gabes.

At the end of this work, I am pleased to express my sincere thanks and gratitude to my supervisor, Mr Béchir CHAOUACHI, director of the Energy, Water, Environment and Processes (LEEEP) research laboratory and professor at the National Engineering School of Gabés (ENIG). I would like to express my deep gratitude for his encouragement, his pertinent suggestions and his advice in the preparation of this work.

I would like to express my deepest gratitude to Mr Ahmed Hannachi, Director of the Doctoral School: Sciences, Engineering and Society and Professor at the National Engineering School of Gabés (ENIG), for his encouragement, his many words of advice and his attentiveness, which were crucial to the success of this project.

Finally, I would like to express my gratitude to all my friends for their help and invaluable advice, and to all the teaching and administrative staff at the ENIG who have contributed in any way to the completion of this work.

I am particularly grateful to my husband, who encouraged me to embark on this journey, and to my family for their encouragement, understanding and often indispensable support along the way.

Summary

As part of this post-doctoral project aimed at continuously improving the performance of absorption-diffusion cycles and in the search for new refrigerants to replace fluids currently in use, we are presenting new hydrocarbon mixtures. We are presenting new hydrocarbon mixtures. The most volatile fluid, propylene, is chosen as the refrigerant, while the least volatile are used as absorbents.

The study of the thermodynamic feasibility of the cycle consists of determining the operating limits of absorption refrigeration machines with the various alkane pairs considered. Modelling using ASPEN PLUS and ASPEN HYSYS flowsheeting software has shown that the proposed hydrocarbon blend gives the best COP compared with those given by the hydrocarbons tested. The calculation of the coefficient of performance will lead to the choice of the best performing blend. In terms of the future of this work, we are proposing to look for other more efficient and environmentally-friendly working combinations that do not have harmful effects on the ozone layer, and to carry out experimental tests on this machine using the propylene/nonane mixture and helium as the inert gas.

Key words: Absorption-diffusion cycles; Hydrocarbon mixture; Coefficient of performance, thermodynamic feasibility, Propene, Alkanes.

GENERAL INTRODUCTION

In this global context, refrigeration machines are an important link in the chain of humanity's fight against this deadly epidemic.

Refrigeration machines are systems that use energy to extract heat from the medium to be cooled (cold source) and release it back to the environment.extract heat from the medium to be cooled (cold and reject it to the outside outside (hot source). Refrigeration machines meet the cooling needs of the domestic, commercial and industrial sectors, i.e. air conditioning, freezing and refrigeration of perishable products.

A refrigeration machine is a closed circuit in which a refrigerant circulates. Its operation is based on the thermodynamic principle, which uses the physical properties of a fluid to transfer heat or energy. The refrigeration cycle comprises four stages: compression, condensation, expansion and evaporation. [1]

The refrigerant is compressed in a compressor, increasing its pressure (high pressure) and temperature (vapour state). The condenser allows the fluid to condense by exchange with an external fluid (water, air, etc.). The refrigerant then returns to its liquid state but retains its high pressure. As it passes through the expansion valve, it is vaporised by a sudden drop in pressure. In the evaporator, the liquid fluid absorbs thermal energy from the medium to be cooled, and a new cycle begins. To characterise the energy efficiency of a refrigeration machine, we use the concept of coefficient of performance, the ratio between the useful refrigeration energy and the energy consumed to produce it.

The way in which the vapours formed in the evaporator are extracted enables several types of machine to be distinguished.

In compression and ejection machines, the method is mechanical. In sorption machines, the vapour is sucked up by attaching it to an absorbent substance that has a high affinity for the refrigerant molecules (example: ammonia refrigerant / water absorbent). [1]

In this type of machines refrigeration the most common, the pure fluid or mixture exists in vapour and liquid states. Conversely, there are systems in which the fluid does not change physical state.

The heat recovery systems and les pompes heat pumps use the same thermodynamic cycle. They are differentiated from refrigeration machines according to whether the emphasis is on heat extraction or heat input [1].

In this global context, the efforts made in scientific, theoretical and experimental research to solve these problems focus on improving the energy performance of vapour compression systems by minimising the quantity of refrigerants used, exploring the use of new fluids with a low environmental impact and using alternative cold production technologies using adsorption and absorption.

Most absorption refrigeration machines run on thermal energy and use either NH_3/H_2O or $LiBr/H_2O$. However, the use of these mixtures suffers from a number of constraints, such as crystallisation in the case of $H_2O/LiBr$ and high pressure in the case of NH_3/H_2O, which leads to toxic ammonia leaks [2].

The performance of these machines can be developed by modelling, which allows a more flexible study of the influence of parameters on the performance of any refrigeration machine based on the absorption principle.

In this context, we are studying the contribution to the study of hydrocarbons as refrigerants in an absorption-diffusion cycle. This is a machine with a moderate cooling capacity.

Mixtures were considered and compared, with the condenser and absorber cooled by ambient air at 35°C. Helium was used as the inert gas. The total operating pressure is around 17.5 bar. The hydrocarbon mixtures were simulated using Aspen Plus and Aspen Hysys flowsheeting software.

This study is structured as follows: The first chapter deals with the design and thermodynamic modelling of an absorption-diffusion refrigeration machine, and the operating principle of the machine has been described. The various operating limits of the mixtures involved were determined using Aspen Plus software.

The second chapter is devoted to interpreting the results obtained by simulation using Aspen Plus and Aspen Hysys software. The theoretical results were validated by comparing them with those in the literature.

The document ends with a general conclusion and a look ahead to the future.

Chapter I

Design, modelling and simulation of a DAR cycle running on hydrocarbons

The aim of this chapter is to study the possibility of using hydrocarbon mixtures as working fluids in absorption-diffusion refrigeration machines. We present new hydrocarbon mixtures, in fact, the most volatile fluid, propylene is chosen as refrigerant, those less volatile, Nonane (n-C9H20), Decane (n-C10H22), undecane (n-C11H24), Dodecane (n-C12H26), Tridecane (n-C13H28), Tetradecane (n-C14H30), Pentadecane (n-C15H32), Hexadecane (n-C16H34) are chosen as absorbents. Table II.1 gives details of the refrigerant/absorbent combinations.

Table.I.1: Hydrocarbon binaries studied

	n-decane n- C10H22	undecane n-C11H24	Dodecane n-C12H26	Tridecane n- C13H28	Tetradecane n- C14H30	Pentadecane n-C15H32	Hexadecane n-C16H34
Propylene (C_3H_6)	+	+	+	+	+	+	+

The study is carried out as follows:

S we study the machine whose absorber and condenser are cooled with ambient air.

S The study of the liquid-vapour equilibrium of the mixture, constituting the working fluid studied, which is represented in the Oldham diagram

S Examination of this diagram can be used to determine the operating temperature range within which the machine can be operated and the possible enrichment of the solution.

1.1. Absorption-diffusion refrigeration machine

1.1.1. Description of an absorption-diffusion refrigeration machine

The absorption-diffusion refrigeration machine was invented by Von Platen et al (1928) [24]. It uses three operating fluids: ammonia (refrigerant), water (absorbent) and hydrogen as an inert gas.

In an absorption-diffusion machine, a supporting gas is used to balance the pressures between the condenser and the evaporator, allowing the refrigerant to evaporate and thus produce cold. Because there are no moving parts in the unit, the absorption-diffusion system is quiet and reliable. It is therefore often used in hotel rooms, offices, and in arid and isolated areas.

1.1.2. Operating principle

The diffusion absorption refrigeration cycle has four main components: a vapour generator (or desorber), a condenser, an evaporator and a vapour absorber.

In a simplified DAR (Diffusion-Absorption Refrigeration cycle.) system cycle shown in Figure (II.2), heat is supplied in the generator directly to the rich solution (Qgen). As can be seen, the refrigerant-absorbent vapour is separated from the rich solution in the generator (2), and the vapour bubbles then rise inside the bubble pump (3). The vapour flows through the condenser, where it condenses, releasing a quantity of heat Q_{Cond}.

The condensate flows directly to the inlet of the evaporator (5). At the evaporator inlet, the liquid refrigerant mixes with the helium and refrigerant residue on arrival at the absorber (7) after passing through the gas-to-gas heat exchanger.

The refrigerant-helium mixture circulates in the co-current flow evaporator. The lean solution enters the absorber from the top after passing through the solution-solution heat exchanger. The refrigerant vapour is absorbed by the lean solution, transforming it into a rich solution that flows into the reservoir (1). The rich solution then leaves the tank for the generator. The helium that is not absorbed continues on its way to the evaporator along with the residue that was not absorbed.

Another absorption-diffusion refrigeration cycle configuration is shown in Figure I.3.

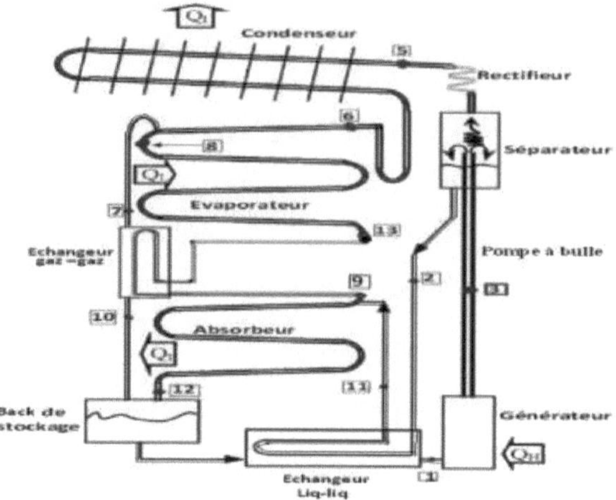

Figure I.1: Schematic diagram of an absorption-diffusion refrigeration cycle [24].

Like every refrigeration system, the DAR system (Figure I.3.) has its advantages and disadvantages:

Advantage:

- It consumes mainly heat energy;

- It can be used if you have a heat source (free heat or waste heat);

- Quiet, vibration-free machines.

Disadvantages:

- The COP is low compared to compression cycles;

- Construction problem; significant waterproofing.

1.2. Thermodynamic models

The choice of the appropriate thermodynamic method for calculating and/or predicting the properties of refrigerant mixtures is of crucial importance for the reliability of the results. For low-pressure systems the two approaches (φ-φ) and (γ-φ) allow this type of calculation to be carried out with good accuracy. For high-pressure systems, the symmetrical method is recommended.

According to previous research work [16-18], the predictions of several models, namely NRTL (Non Random Two Liquids), the SRK (Soave-Redlich-Kwong) equation and the PC-SAFT (Perturbed-Chain Statistical Associating Fluid Theory) equation, have been compared with experimental liquid-vapour equilibrium data for several hydrocarbon mixtures. This study was carried out by Chekir et al [16,17]. In our work, we are interested in the Peng-Robinson and Chao-Seader models and the PC-SAFT model [18,19]. In a previous study, Chekir et al [20] showed that, at low pressure, the three models reproduce the experimental data fairly well.

A recent study by Dardour et al [18] showed that, compared with the Chao-Seader model, the Peng-Robinson model represented the liquid-vapour equilibrium of all the isothermal and isobaric curves for the alkane mixtures considered, the Peng-Robinson model represented the thermodynamic properties of these mixtures more faithfully, particularly at the level of the boiling curves. The authors also showed that the Peng-Robinson model was more faithful to the experimental data than the Chao-Seader model.

1.2.1. Calculating liquid-vapour equilibrium :

For a two-phase fluid system, liquid-vapour, with n constituents, we have [18] :

$$d(nG)^L = -(nS)^L dT + (nV)^L dP + \sum_i \mu_i^L dn_i^L \tag{I.1}$$

$$d(nG)^v = -(nS)^V dT + (nV)^V dP + \sum_i \mu_i^V dn_i^V \tag{I.2}$$

S, V and μ being the molar entropy, the molar volume and the chemical potential. The exponents

(L) and (V) refer to the liquid and vapour phase respectively.

The change in free enthalpy of the two-phase system is then :

$$d(nG) = d(nG)^L + d(nG)^V \tag{I.3}$$

The result is :

$$d(nG) = (nV)dP - (nS)dT + \sum_i \mu_i^L dn_i^L + \sum_i \mu_i^V dn_i^V \tag{I.4}$$

The criterion for phase equilibrium of the system is, at fixed P and T, given by :

$$d(nG) = 0 \qquad (II.5)$$

Either,

$$\sum_i \mu_i^L \, dn_i^L + \sum_i \mu_i^V \, dn_i^V = 0 \qquad (II.6)$$

For a closed system and in the absence of chemical reactions we have: dnf = -dn^.

Equation II.6 becomes :

$$\sum_i (\mu_i^L - \mu_i^V) \, dn_i^L = 0 \qquad (II.7)$$

Hence finally

$$\mu_i^L = \mu_i^V \qquad (II.8)$$

By introducing the definition of fugacity f and using the same standard state for both phases, equation (II.8) becomes :

$$\mu_i^\circ + RT \ln \frac{f_i^L}{f_i} = \mu_i^\circ + RT \ln \frac{f_i^V}{f_i} \qquad (II.9)$$

Hence

$$f_i^L = f_i^V \qquad (II.10)$$

According to Gibbs' phase rule, the binary system in liquid-vapour equilibrium is bivariant:

$$v = c + 2 - \varphi = 2 + 2 - 2 = 2 \qquad (II.11)$$

and φ are the number of components and phases respectively.

To characterise the state of such a system, all we need to do is set the values of the two intensive parameters. In fact, for such a system we can write successively :

$$\begin{cases} f_1^L(T,P,x_1,x_2) = f_1^V(T,P,y_1,y_2) \\ f_2^L(T,P,x_1,x_2) = f_2^V(T,P,y_1,y_2) \\ \quad y_1 + y_2 = 1 \\ \quad x_1 + x_2 = 1 \end{cases} \qquad (II.12)$$

At a fixed T and P, the system consists of 4 equations with 4 unknowns: $y1\_\_y_2$ $x1 and$-x_2 . It is therefore sufficient to know two of these unknowns to be able to solve the system and determine the rest of the unknown state variables. To calculate the equilibrium, we need to calculate the liquid and vapour fugacities of the species present. There are two methods for doing this, one of which, called $(\emptyset - \emptyset)$, is symmetrical, while the other, called $(\gamma - \emptyset)$, is asymmetrical.

> **Approach** $(\emptyset - \emptyset)$

The symmetrical approach to calculating thermodynamic equilibrium uses an equation of state that describes both the liquid and vapour phases of the fluid, and from which the fugacities can be calculated.

$$f_i^L = \emptyset_i^L x_i P \tag{I.13}$$

$$f_i^V = \emptyset_i^V y_i P \tag{I.14}$$

0L and 0V are the fugacity coefficients of component i in the liquid and vapour phase respectively. They depend on temperature, pressure and composition.

The phase equilibrium is then written :

$$\emptyset_i^L x_i = \emptyset_i^V y_i \tag{I.15}$$

> **Approach** $(\gamma - \emptyset)$:

This so-called classical method of calculation uses an equation of state for the vapour phase and an often empirical model of the activity coefficients in the liquid. The fugacities of the species in the vapour phase are expressed as described above (II.13):

$$f_i^V = \emptyset_i^V y_i P \tag{I.16}$$

And that in liquid phase are given by :

$$f_i^L = \gamma_i^L x_i f_i^{*L} \tag{I.17}$$

YL is the activity coefficient of species i in the liquid phase, f_i*L is the liquid fugacity of species i at the pressure and temperature of the mixture.

$$f_i^{*L}(T,P) = P_i^{sat}\,\emptyset_i^*(T,P_i^{sat})\exp\left(\frac{v_i^L(P-P_i^{sat})}{RT}\right) \tag{I.18}$$

V^{b} being the molar volume of species i in the liquid phase in the saturated state.

Equation II.10 becomes :

$$\gamma_i^L x_i f_i^{*L} = \emptyset_i^V y_i P \tag{I.19}$$

Advantages and disadvantages of each approach [20,21].

- $(\emptyset - \emptyset)$ **approach**: thermodynamic calculations using the symmetrical $(\emptyset - \emptyset)$ approach are reliable over wide pressure and temperature ranges, including subcritical and supercritical regions. For ideal or slightly non-ideal systems, the thermodynamic properties, whether in the vapour or liquid phase, can be determined with a minimum of data on the constituents. For a correct and good representation of non-ideal systems it is recommended to start by determining the binary interaction parameters by regression of the experimental liquid-vapour equilibrium data.

The use of cubic equations of state generally does not allow liquid densities to be calculated with sufficient accuracy. In this case, it is recommended that an additional calculation model specific to this application be added to the equation of state.

- $(\gamma - \emptyset)$ **approach:** the asymmetric method $(\gamma - \emptyset)$ is the best method for describing liquid mixtures with a very pronounced non-ideality.

- However, it is necessary to determine the binary parameters from the experimental data. The asymmetric method should only be used for low-pressure systems.

- **Equilibrium constants [15]:** Their values measure the proportions in which a species i is shared between the two liquid and vapour phases in thermodynamic equilibrium. For each species, the equilibrium constant is written as :

$$K_i = \frac{y_i}{x_i} = \frac{\emptyset_i^L}{\emptyset_i^V} \tag{I.20}$$

When the symmetrical method $(\emptyset - \emptyset)$ is used.

Similarly Ki is given by:

$$K_i = \frac{y_i}{x_i} = \frac{\gamma_i^L f_i^{\circ L}}{\phi_i^V}$$
(I.21)

If we adopt the asymmetric method . $(\gamma - \phi)$

1.2.2. Peng-Robinson model

In the 1970s, doctoral student D. Peng and his thesis supervisor Professor D. B. Robinson of the University of Alberta (Edmonton, Canada) developed the equation of state that has since borne their names. The main objectives of their work were to produce an equation of state suitable for natural gas processes, including [18] :

- The parameters depend solely on the critical properties and the acentric factor of the constituents,

- The application remains reliable and valid in the vicinity of the critical point, particularly for calculating the compressibility factor and liquid density,

- The mixing rules require only one binary interaction parameter, independent of temperature, pressure and composition.

The Peng-Robinson thermodynamic model uses [17] :

- The Peng-Robinson cubic equation of state for all thermodynamic properties except liquid molar volume [18].

- The API method for calculating the liquid molar volume of pseudo-compounds and the Rackett model for real compounds.

The PENG-ROBINSON method gives satisfactory results whatever the temperature and pressure values. The method is consistent in the critical region: it does not display abnormal behaviour like methods using activity coefficients. The results therefore remain correct and accurate in the vicinity of the critical mixing point [15].

Peng-Robinson equation of state

> Case of a pure body

The Peng-Robinson equation of state is written for a pure body whose state is defined by the triplet *(P*: pressure, *V*: molar volume, *T*: temperature), in the form: [2].

$$P = \frac{RT}{V-b} - \frac{a\alpha}{V^2+2bV-b^2}$$

(I.22)

With

$$a = 0{,}45724\frac{R^2T_c^2}{P_c}\,\alpha$$
$$b = 0{,}0778\frac{RT_c}{P_c}$$

(I.23)

$$\alpha = \left[1 + (0{,}37464 + 1{,}5422w - 0{,}26992w^2)\left(1 - \sqrt{\frac{T}{T_c}}\right)\right]^2$$

T_c , P_c and w are the critical temperature, critical pressure and acentric factor respectively.

As this is a cubic equation of state in volume, the compressibility factor Z is a solution of the following cubic equation :

$$Z^3 - (1 - B)Z^2 + (A - 3B^2 - 2B)Z - (AB - B^2 - B^3) = 0$$

(I.24)

With

$$\begin{cases} A = \frac{A\alpha P}{R^2T^2} = 0{,}45724\frac{\alpha P_r}{T_r^2} \\ B = \frac{bP}{RT} = 0{,}07780\frac{P_r}{T_r} \end{cases}$$

(I.25)

T_r and P_r are the reduced temperature and reduced pressure respectively $\left(T_r = \frac{T}{T_c}\ etP_r = \frac{P}{P_r}\right).$

> **Case of a mixture**

In this case, the equation of state is written as [18] for a system whose state is defined by the triplet (P,V,z) *(P*: pressure, *V*: molar volume, T: temperature with z the overall composition):

$$P(T,V,z) = \frac{RT}{V-b_m} - \frac{(a\alpha)_m}{V^2+2b_mV-b_m^2}$$

(I.26)

$(a\,\alpha)_m etb_m$ are estimated using the VAN DER WAALS mixing rules:

$$\begin{cases} (a\,\alpha)_m = \sum_{i=1}^{nc}\sum_{j=1}^{nc} z_iz_j(a\,\alpha)_{ij} = \sqrt{(a\,\alpha)_i(a\,\alpha)_j}(1 - k_{ij}) \\ b_m = \sum_{i=1}^{nc} z_iB_i \end{cases}$$

(I.27)

For all binaries we take :
$$k_{ij} = k_{ji} \text{ et } k_{ii} = k_{jj} = 0 \qquad \text{(I.28)}$$

> Procedure for calculating binary liquid-vapour equilibrium

The mole fractions in the liquid and vapour phase, x and y, of two components i and j are related by equation [15] :

$$y_i = K_i x_i \quad (i \in \{1, 2\}) \qquad \text{(I.29)}$$

Or

K_i is the equilibrium constant which can be determined from equation :

$$K_i = \frac{\emptyset_i^l}{\emptyset_i^v} \qquad \text{(I.30)}$$

And $\emptyset_i^l$ et $\emptyset_i^v$, the fugacity coefficients of component i in the liquid and vapour phase respectively.

The fugacity coefficient of component i in the vapour phase $0^{\wedge}$ is calculated from equation :

$$\ln(\emptyset_i^v) = \frac{b_i}{b_m}(Z_v - 1) - \ln(Z_v - B) - \frac{A}{2\sqrt{2}B}\left(\frac{2}{(a\alpha_m)}\Sigma_j y_i(a \, \alpha_{ij}) - \frac{b_i}{b_m}\right) \ln\left(\frac{Z_v+(\sqrt{2}+1)B}{Z_v-(\sqrt{2}-1)B}\right) \qquad \text{(I.31)}$$

0- is determined from a similar equation by replacing the vapour phase compressibility factor Z_v by that for the liquid phase/;.

1.3. Variance and equation of the various machine components

The classic generator of the absorption-diffusion machine shown in the following figure (figure I.4.) consists of a boiler and a bubble pump. The rich solution (12) leaving the absorber and heated by the lean solution (2) feeds the inner tube or tubes of the generator acting as a bubble pump. Under the effect of the driving heat supplied to the generator, vapour bubbles form inside each tube [2].

In the bag mode, these vapour bubbles act as pistons which carry the solution up the tube to flow into the separator, which acts as a phase separator [24]. The lean solution thus obtained (2) is sent to the absorber. The vapour bubbles and the vapour produced by the boiler accumulate in the separator at the outlet of the generator (3) and flow to the rectifier above. The ammonia/water/helium system is considered for the variance calculation.

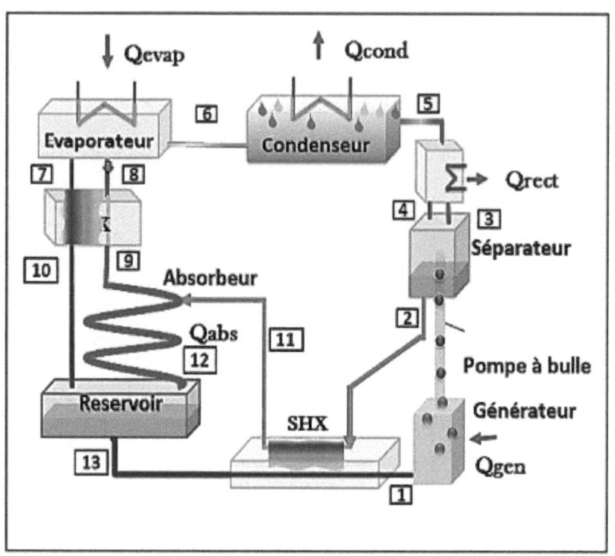

Legend:

Qcond: Power output at condenser (W)

Qrect: Heat released at rectifier level (W)

Qgen: Driving power supplied to the generator (W)

Qevap: Cooling capacity (W)

Qabs: Power released at the absorber

GHX: Gas-to-gas exchanger

SHX : Solution exchanger

Figure.I.2: Schematic diagram of an absorption-diffusion machine [2].

1.3.1. Generator

The generator has four material flows, two incoming, two outgoing, and one energy flow:

- Flow 1: binary mixture of water and ammonia - var = 3. $\qquad$ (I.32)

- Flow 2: binary mixture of water and ammonia - var = 3 $\qquad$ (I.33)

- Flow 3: binary mixture of water and ammonia - var =3 $\qquad$ (I.34)

-Flow 4: water - var = 2 $\qquad$ (I.35)

22

-Energy flow QG - var = 1 $\qquad$ (I.36)

The conservation of matter and energy equations governing the generator are written as follows:

Overall material balance :

m3+m2=m1+m4 $\qquad$ (I.37)

Partial report on ammonia :

m3X3+m2Y2 =m1X1 $\qquad$ (I.38)

Energy balance :

$m_2h_2 + m_3h_3 - m_1h_1 - m_4h_4 = QGEN$ $\qquad$ (I.39)

The variance of the generator is then equal to :

VGEN = ii+i- 3= 9 $\qquad$ (I.40)

Taking into account that flow 2 is the lean solution which is a liquid in a saturated state and that flow 3 is a vapour which leaves the generator in a saturated state, the variance of the generator becomes equal to 7.

1.3.2. Grinder

The rectifier shown in the figure above has 3 flows: one incoming and two outgoing, plus an energy flow:

- Flow 3: binary mixture of water and ammonia - var = 3 (I.4i)

- Flow 5: subcooled ammonia - var = 2 $\qquad$ (I.42)

- Flow 4: water - var = 2.

- Energy flow. QRect - var = i. $\qquad$ (I.43)

The conservation of matter and energy equations governing the rectifier are written as follows:

Overall material balance :

$m5 + m4 = m3$ (I.44)

Partial report on ammonia :

$m3\,Y3 = m5$ (I.45)

Energy balance :

$m3h3-m5h5 -m4h4=QRect$ (I.46)

The variance of the rectifier is then equal to :

$VRect = 7+i-3= 5$ (I.47)

Taking into account that the steam (5) is in a saturated state and that the reflux (4) is a liquid in a saturated state, the variance of the rectifier becomes equal to 3.

1.3.3. Condenser

The condenser shown in the previous figure has two material flows, one incoming and one outgoing, and one energy flow:

- Flux 5: ammonia - var $= 2$.   (I.48)
- Flow 6: subcooled ammonia - var $= 2$ (I.49)
- Energy flow: QCond - var $= 1$ (I.50)

The conservation of matter and energy equations governing the condenser are written as follows:

Overall material balance :

$m5 = m6$ (I.51)

Energy balance :

$m5\,h5 -m6\,h6 = QCond$ (I.52)

The variance of the condenser is then equal to :

$4+1-2=3$ (I.53)

Taking into account that the steam (5) is in a saturated state, and the flow (6) is a subcooled liquid, the variance of the condenser becomes equal to 1.

1.3.4. Evaporator

The evaporator has 3 material flows, one outgoing and two incoming, and one energy flow:

- Flow 6: ammonia -var = 2   (I.54)

- Flux 7: ammonia + helium - var = 3   (I.55)

- Flux 8: ammonia + helium - var = 3   (I.56)

- Energy flow: QEvap -var= 1   (I.57)

The conservation of matter and energy equations governing the evaporator are written as follows:

Overall material balance :

$m7 + m6 = m8$ (I.58)

Partial report on ammonia :

$m6Y6 + m7Y7 = m8Y8$ (I.59)

Partial balance on helium :

$m6Y6he + m7 Y7he = m8 Y8he$ (I.60)

Energy balance :

$m8h8 - m6h6 - m7h7 = QEvap$ (I.61)

The variance of the evaporator is then equal to :

$8 + 1 - 4 = 5$ (I.62)

1.3.5. Gas - gas exchanger

The gas-gas exchanger is traversed by 4 material flows, two incoming and two outgoing:

- Flux 7: ammonia + helium - var = 3   (I.63)

- Flux 8: ammonia + helium - var = 3   (I.64)

- Flux 9: ammonia + helium -var =3   (I.65)

- Flux 10: ammonia + helium - var = 3   (I.66)

The conservation of matter and energy equations governing the condenser are written as follows:

Overall material balance :

$$m_9 + m_7 = m_{10} + m_8 \quad (I.67)$$

Partial report on ammonia :

$$m_9 Y_9 + m_7 Y_7 = m_{10} Y_{10} + m_8 Y_8 \quad (I.68)$$

Partial balance on helium :

$$m_9 Y_{9he} + m_7 Y_{7he} = m_{10} Y_{10he} + m_8 Y_{8he} \quad (I.69)$$

Energy balance :

$$m_9 h_9 + m_7 h_7 - m_s h_8 - m_{io} h_{io} = Q_{EG} \quad (I.70)$$

The variance of the gas-to-gas heat exchanger is then equal to

$$i2 + i - 4 = 9 \quad (I.7i)$$

1.3.6. Absorber

The absorber has 3 material flows, two incoming and one outgoing, and one energy flow:

- Flux9 : ammonia + helium - var $= 3$ (I.72)

- Flow ii: ammonia + water - var $= 3$ (I.73)

- Fluxi2: ammonia + water - var $= 3$ (I.74)

- Energy flow: QAbs - var $= i$ (I.75)

The conservation of matter and energy equations governing the absorber are written as follows:

Overall material balance :

$$m_9 + m_{ii} = m_{i2} \quad (I.76)$$

Partial report on ammonia :

$$m_9 Y_9 + m_{ii} X_{ii} = m_{i2} X_{i2} \quad (I.77)$$

Partial balance on helium :

$$m_9 Y_{9he} = m_{i2} Y_{i2he} \quad (I.78)$$

Energy balance :

$$m_9 h_9 + m_{ii} h_{ii} - m_{i2} h_{i2} = Q_{abs} \quad (I.79)$$

The variance of the absorber is then equal to :

$9+i-4 =6$ (I.8o)

1.3.7. Solution exchanger

The solution exchanger is traversed by 4 material flows, two incoming and two outgoing:

-     Flux i: ammonia + water - var = 3   (I.8i)

-Flow 2: ammonia + water - var = 3 (I.82)

-     Flux 13: ammonia + water - var = 3   (I.83)

-     Flow 11 =Flow 2: ammonia + water - var = 3   (I.84)

The conservation of matter and energy equations governing the solution exchanger are written

as follows:

Overall material balance :

$m1 = m_{12-m10}$ (I.85)

Partial report on ammonia :

$m1X1 = m_{12}X_{12} - m_{10}X_{10}$ (I.86)

Energy balance :

$m1h1 - m_{12\,h12+m10\,h10} = QES$ (I.87)

The variance of the solution exchanger is then equal to :

$12+1-3=10.$ (I.88)

Then the cycle variance is equal to :

$7+3+1+5+9+6+10= 41.$ (I.89)

The variables that are counted twice and must be subtracted from the cycle variance are

shown in the following Table I.2 :

Table.I.2: Number of repeated variables

	Flow	Variance
Generator-Rectifier	Flow 3	3
	Flux 4	2

Rectifier-Condenser	Flow 5	2
Condenser-Evaporator	Flow 6	2
Evaporator-Gas exchanger-	Flow 7	3
gas	Flow 8	3
Gas-gas exchanger -Absorber	Flux 10	3
	Flow 9	3
Absorber - Solution exchanger	Flux 11	3
	Flow 13	3
Solution-Generator exchanger	Stream 1	3
	Flux 2	3
TOTAL		33

Hence the total cycle variance becomes , taking into account the total cycle pressure :

Vcycle = 41+1-33= 9. (I.90)

We therefore need to take 9 independent variables

1.4. Design and modelling of the machine using Aspen Plus software

This section presents the simulation tool used, ASPEN PLUS (Advanced System for Process Engineering) on the absorption-diffusion machine, and the various results of the study of the thermodynamic feasibility of hydrocarbon binaries.

1.4.1. Presentation of the simulation tool

ASPEN PLUS is one of the best-known flowsheeting software packages [30]. It can be used to simulate existing processes in a number of sectors, including food processing, mineral processing and petrochemicals.

Developed and marketed by ASPEN TECHNOLOGY, it is designed for engineers, managers and researchers.

The main features of this software are :

> Calculation and prediction of thermodynamic properties.

> Smoothing of experimental data and establishment of analytical correlations.

> Simulation, sizing, estimation and economic evaluation and optimisation of equipment and

processes.

To perform these functions, the software has :

> A library of modules for calculating the physical and thermodynamic properties of pure

substances and mixtures.

> A library of numerical calculation methods for solving algebraic and differential systems.

> A library of standard modules for simulating equipment (exchangers, reactors, columns,

etc.).

1.4.2. Description of the refrigeration machine

The basic components of the machine are the generator, rectifier, condenser, evaporator,

absorber, gas-gas exchanger and solution exchanger.

1.4.3. Modelling using ASPEN Plus software

1.4.3.1. Fundamental data

They are determined by the system's cooling capacity and its maximum theoretical COP. The

proposed installation has a cooling capacity of 300 W.

The ambient air has an average temperature of 35°C. Set the operating conditions for the main

machine components as follows:

> a refrigerant evaporator outlet temperature equal to 0°C.

> Condenser: In the case of cooling with air, the temperature at the end of condensation is

assumed to be 12 to 15°C higher than that of the air, i.e. between 47 and 50°C. The liquid,

assumed to be undercooled by 5°C, leaves the condenser at 42°C or 45°C.

> Absorber: The same reasoning as for the condenser applies to the absorber, since their

heat is generally evacuated to the same heat sinks.

> Generator: The driving heat supplied to the generator can come from several sources: solar collectors, thermal waste, steam, combustion gases.

1.4.3.2. Configurations of the various components of the diffusion absorption machine

The Aspen Plus software contains many models that can represent the machine components. The table below shows the different configurations chosen for this simulation:

Table.I.3: Different configurations used for the simulation

Component	Aspen model
Generator	Flash2
Grinder	Heater
Condenser	Heater
Evaporator	Heater
Absorber	Radfrac
Solution exchanger	Heatx
Gas-gas exchanger	Heatx

The machine simulation diagram is shown in the following figure:

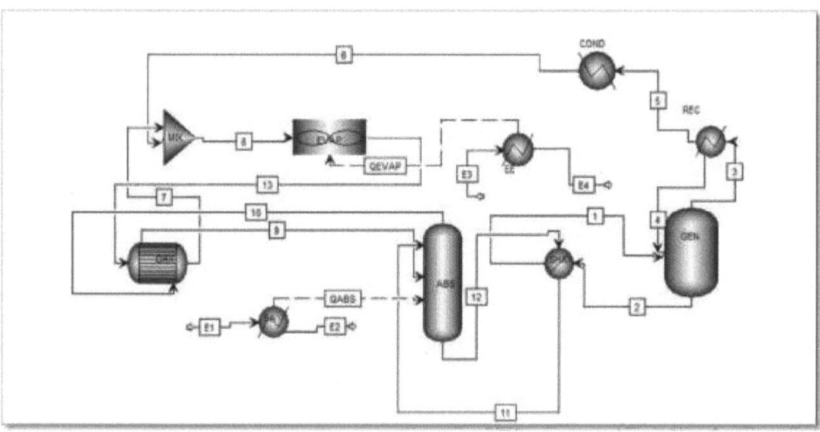

Figure.I.3: Flowsheet diagram of the absorption-diffusion refrigeration cycle of the

30

machine.

1.4.3.3.   Assumptions

> -The machine is assumed to operate in steady state and pressure losses are considered negligible [18,20].

> -The purity of the refrigerant vapour leaving the generator is between 97 and 99.8%.

> -The refrigerant vapour leaving the generator towards the condenser is assumed to be saturated. The lean solution at the generator outlet is saturated at the driving temperature of the cycle.

> -The two heat exchangers are assumed to be adiabatic and each characterised by a thermal pinch; that of the exchanger (gas/gas) is set at 10°C and that of the solution exchanger (liquid/liquid) is set at 12°C.

The 9 variables to be identified, based on operating assumptions and technical data, are detailed in Table I.4.

Table.I.4: Data and working hypotheses.

Data and assumptions	Value
Generator temperature	90-180°C
Total pressure	17.16 bar
Condenser temperature	42°C
Evaporator outlet temperature	0°C
Absorber outlet temperature	30°C
Molar composition of the rich solution	0,39
Machine cooling capacity	300 W
Temperature pinch in the steam-steam exchanger	10°C
Temperature pinch in the solution exchanger	12°C

1.5. Use of propylene (C3H6) as a refrigerant

This part is devoted to the study of the thermodynamic feasibility of the cycle, which consists of determining the operating limits of absorption refrigeration machines with various mixtures of alkanes in compliance with the operating conditions already set.

The alkane pairs considered are C3H6/n-C10H22, C3H6/n-C11H24, C3H6/n-C12H26, C3H6/n-C14H30, C3H6/n-C15H32 and C3H6/n-C16H34.

- **Cycle layout**

Given the operating conditions set (Table II.4), the vaporisation and condensation temperatures are equal to 0°C and 42°C respectively.

These two temperatures, together with the purity of the refrigerant (equal to 99.6%), determine its condensation and vaporisation pressures. The steps for tracing the machine cycle in the Oldham diagram for each hydrocarbon binary are :

The two temperatures, vaporisation and condensation, are marked on the diagram and the verticals are raised until they intersect the iso-title curves corresponding to 99.6% of the refrigerant. The points of intersection represent the evaporator and the condenser and allow the values of high and low pressure (P_evap; P_cond) to be read directly from the diagram. The assumptions adopted are :

- There is a balance between the different devices,
- The pressures at the evaporator and absorber are equal (low pressure) and the pressures at the generator and condenser are equal (high pressure).

1.6. Conclusion

In this chapter, we have proposed as working fluids a binary mixture of hydrocarbons that are rarely used today in diffusion absorption refrigeration machines.

Firstly, we described the operation of our installation, followed by thermodynamic modelling, and then we determined the variance of our installation. The material and energy balance equations and the liquid-vapour equilibria for the various mixtures are then determined. Finally, we presented several thermodynamic models in order to choose the most suitable model for our installation, which is the PENG ROBINSON model. By plotting the Oldham diagram for the various binaries considered, we assessed the thermodynamic feasibility of using these hydrocarbon pairs by determining the operating limits for each binary. To do this, we have complied with the operating conditions initially set, such as steady-state operation of the machine, negligible pressure drops and 99.6% purity of refrigerant vapour at the generator outlet.

Simulation and use of results

In this chapter, we will detail the results of the plant simulation based on the thermodynamic study presented in the previous chapter. We present our investigations into hydrocarbon mixtures for the absorption-diffusion refrigeration cycle. We begin by modelling the various components of the machine and then present the detailed simulation results, followed by a discussion. The calculation of the coefficient of performance will lead to the choice of the most efficient mixture, which will be dealt with in detail by means of a parametric study. Our work will be validated by comparison with recent experimental and theoretical results in the literature.

11.1. Design of the various components of the diffusion absorption machine

The Aspen Hysys software contains many models that can represent the machine components. Table II.1 shows the different models chosen for this simulation. The different devices are designed as follows:

> GEN generator

In the flowsheeting of our machine, the generator is presented by the RADFRAC model (Figure II.1). This is fed by the preheated solution from the absorber and has two outputs corresponding to the vapour and liquid streams.

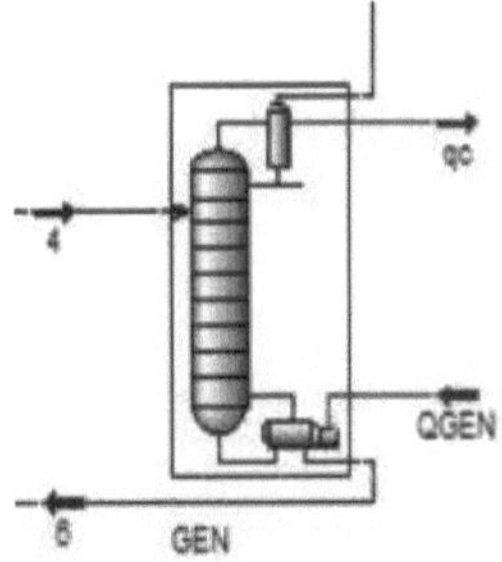

Figure II.1: Generator diagram

> Condenser COND

The condenser is presented by a HEATER heat exchanger model schematised in Figure II.2 below:

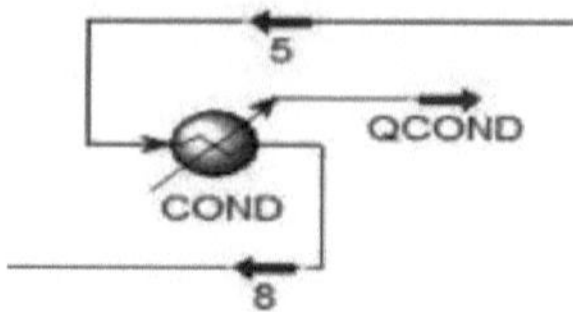

Figure II.2: Condenser diagram

> EVAP evaporator

The evaporator consists of two blocks, a HEATER and a MIXER, shown in figure.II.3, below:

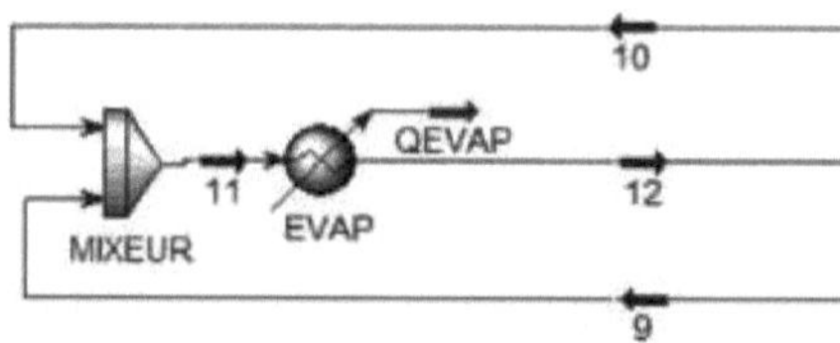

Figure II.3: Diagram of the evaporator

> ABS absorber

The absorber consists of three blocks, a HEATER, a MIXER and a COOLER, shown in Figure II.4 below:

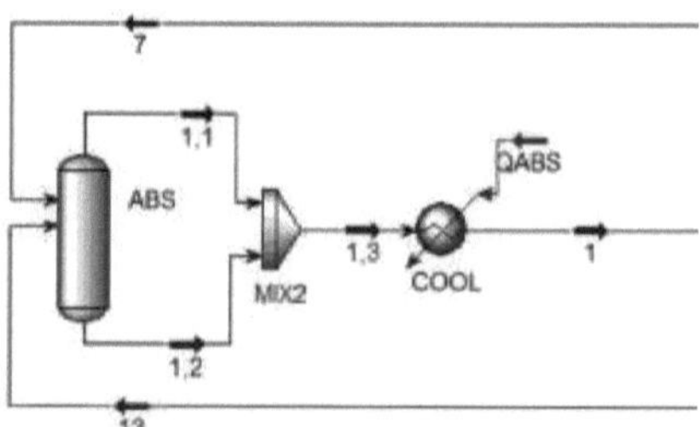

Figure II.4. II.4 Schematic diagram of the absorber

> SHX Solution exchanger

The solution exchanger is presented by two HEATER heat exchanger blocks shown in Figure II.5, below:

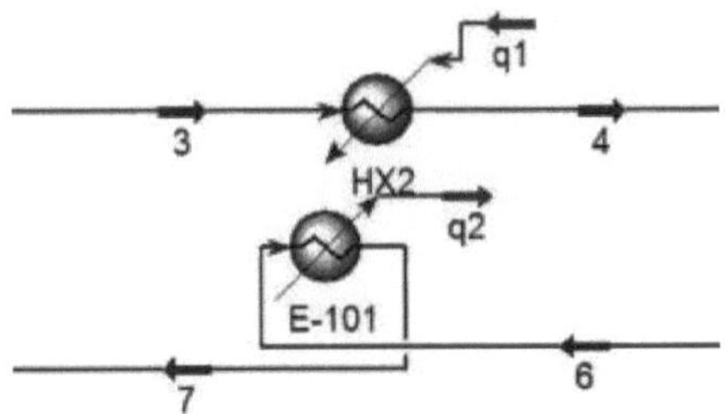

Figure III.5 Solution exchanger diagram

> GHX gas exchanger

The gas exchanger is presented by a HEATEX heat exchanger model, which is shown in Figure II.6 below:

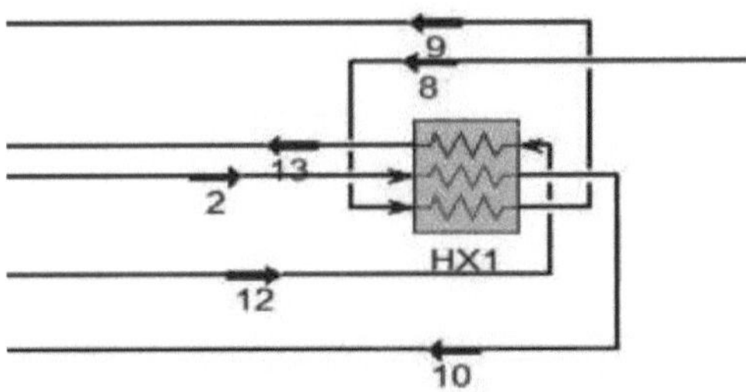

Figure II.6. II.6 Schematic diagram of the gas exchanger

The models of the various components of the absorption-diffusion machine under flowsheeting are illustrated in the following table:

Table II.1: Model of the different components in Aspen Hysys

Component	Aspen model

Generator	Radfrac
Condenser	Heater
Evaporator	Heater
Absorber	flash
Solution exchanger	Heater
Gas-gas exchanger	Heatx

The diffusion absorption machine under the Aspen Hysys interface is shown in Figure (II.7) below:

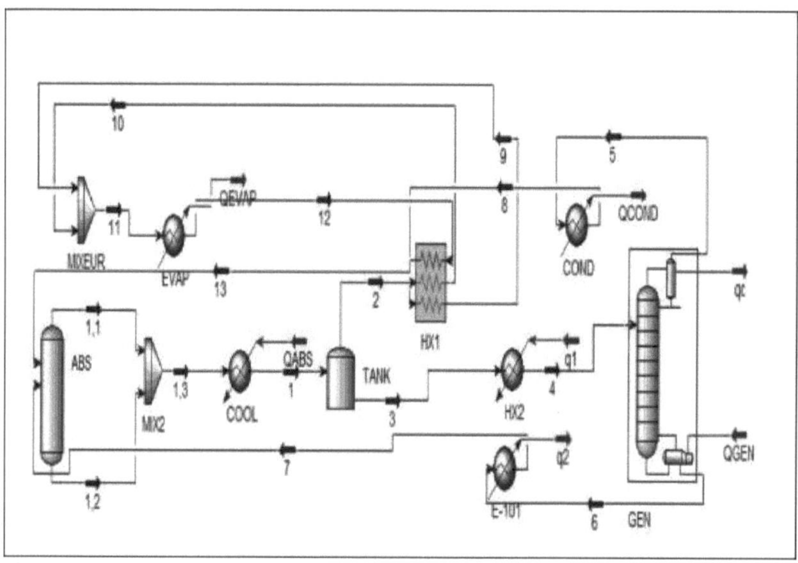

Figure.II.7: Schematic representation of the absorption-diffusion machine in ASPEN-

HYSYS

11.2. Using the results

Taking into account the assumptions and operating conditions mentioned in Table (II.2), the results of the simulation of the absorption-diffusion cycle operating with the different hydrocarbon binaries are presented below. The combinations of hydrocarbon binaries are as follows:

The most volatile fluids, propylene, chosen as the refrigerant, and the least volatile, nonane (n-C_9H_{20}), decane (n-$C_{10}H_{22}$), undecane (n-$C_{11}H_{24}$), dodecane (n-$C_{12}H_{26}$), tridecane (n-$C_{13}H_{28}$), tetradecane (n-$C_{14}H_{30}$), pentadecane (n-$C_{15}H_{32}$) and hexadecane (n-$C16H34$), are used as absorbents. As a reference binary, we simulate the machine with the ammonia/water system, under the same operating conditions.

Table II.2: Data and working hypotheses.

Data and assumptions	Value
Generator temperature	90-230° C
Total pressure	17.16 bar
Condenser temperature	42°C
Evaporator outlet temperature	-10°C - 2 °C
Absorber outlet temperature	30°C
Molar composition of the rich solution	0.35 -0.55
Machine cooling capacity	300watt
Temperature pinch in the liquid-steam exchanger	10°C
Temperature pinch in the solution exchanger	12°C

- Calculation of coefficient of performance [84- 86]

The cycle followed by the refrigerant-absorber pair is called trithermal in the sense that the latter exchanges heat with three heat sources: the first is at the evaporator, the second, known as intermediate, is at the absorber and the condenser, and the last, representing the hot source, is at the generator. The COP is defined by the ratio of the power extracted at the evaporator $Q_{évap}$ and the thermal power supplied by a burner to the generator $Q_{gén}$:

$$COP = Q_{évap}/Q_{gén} \qquad (II.1)$$

II.2.1. Machine operating with Propene as refrigerant

In this section, we present all the figures illustrating the flowsheet. The simulation results in terms of molar compositions, molar flow rates, temperature and thermal power exchanged by the machine for each hydrocarbon binary studied are given in the corresponding tables.

Under the same operating conditions, with a cooling capacity of 300 watts, a total pressure of 17.5 bar and an evaporator temperature of 0°C, an absorption-diffusion machine operating with the propylene/n-nonane mixture gives the best coefficient of performance, a value equal to 0.2231, a value comparable to that found for the ammonia/water system (COP = 0.2504), under the same operating conditions.

In the case of an installation operating with propane as the refrigerant and for a cooling capacity of 1 kW, a total pressure of around 17.5 bar and an evaporator temperature of around 2°C, the propane/n-nonane binary gives the best coefficient of performance, which is 0.2632. The use of butane as a refrigerant leads to a significantly low coefficient of performance of the order of 0.1254. This value is comparable to that found by Ben Ezzine et al. This value is comparable to that found by Ben Ezzine et al [47], under the same operating conditions, which validates our results.

II.3. Conclusion

During this chapter, the use of Aspen Hysys software enabled us to carry out a detailed absorption-diffusion cycle study for each hydrocarbon binary and to search for the best-performing mixture for the operating conditions set. The results show that propane/n-nonane is the best-performing mixture. This work should be followed up by an experimental study to validate the theoretical results and make the necessary improvements.

General conclusion and outlook

This work is a continuation of the work carried out as part of my doctoral thesis, in which a contribution was made to the study of hydrocarbons as alternative mixtures to the couples currently used in absorption-diffusion cycles. Using a machine with a low cooling capacity (300 W), we theoretically studied the binary pairs propylene, propane and butane as refrigerants and the light hydrocarbons hexane, heptane, octane and nonane as absorbents.

This work involves the search for other, more efficient and environmentally-friendly working couples that do not have harmful effects on the ozone layer. In addition, as part of the development of absorption-diffusion refrigeration machines and the search for new refrigerants to replace the fluids currently in use, we are presenting new hydrocarbon mixtures. The most volatile fluid, propylene, has been chosen as the refrigerant, while the least volatile, n-Nonane (n-C9H20), Decane (n-C10H22), undecane (n-C11H24), Dodecane (n-C12H26), Tridecane (n-C13H28), Tetradecane (n-C14H30), Pentadecane (n-C15H32) and Hexadecane (n-C16H34) are chosen as absorbents.

This study is structured as follows: The first chapter deals with the design and thermodynamic modelling of an absorption-diffusion refrigeration machine and the description of its operating principle.

Firstly, we described the operation of our installation, followed by thermodynamic modelling, and then we determined the variance of our installation. The material and energy balance equations and the liquid-vapour equilibria for the various mixtures are then determined.

Finally, we presented several thermodynamic models in order to choose the most suitable model for our installation, which is the PENG ROBINSON model. By plotting the Oldham diagram for the various binaries considered, we assessed the thermodynamic feasibility of using these hydrocarbon pairs by determining the operating limits for each binary. To do this, we have respected the operating conditions initially set, such as steady-state operation of the machine, negligible pressure drops, and the purity of the refrigerant vapour at the generator outlet is taken to be 99.6%.

The final part is devoted to a study of the thermodynamic feasibility of the cycle. This involved determining the operating limits of absorption refrigeration machines with the various alkane pairs considered, using Aspen Plus software.

The second chapter is devoted to the interpretation of the results obtained by simulation using Aspen Plus and Aspen Hysys software. Using the Aspen Hysys software, we were able to carry out a detailed study of the absorption-diffusion cycle for each hydrocarbon binary and to look for the best-performing mixture for the set operating conditions, which in this case is propane/n-nonane.

Based on the thermodynamic study presented in the previous chapter, we modelled the various components of the machine and then presented the detailed simulation results, together with a discussion.

The calculation of the coefficient of performance will lead to the choice of the best performing mixture, which has been dealt with in detail by means of a detailed parametric study. Our work was validated by comparison with recent experimental and theoretical results in the literature.

As a follow-up to this work, and in order to further improve the performance of these absorption-diffusion cycles, we are proposing to look for other more efficient and environmentally-friendly working couples that do not have harmful effects on the ozone layer,

References

[1] engineering book ,https://www.techniques-ingenieur.fr/glossaire/machine- refrigeration

[2] Soli N., Zrelli A., Chaouachi B., Contribution to the study of a diffusion absorption cycle operating with a mixture of hydrocarbons ,2018.

[3] Overview of the History of Cold Production, INTERNATIONAL INSTITUTE OF REFRIGERATION,

[4] Srikhirin P., Aphornratana S., Chungpaibulpatana S. A review of absorption refrigeration technologies. Renewable & Sustainable Energy Reviews, Vol. 5, 2001, 343-372.

[5] Wang R., Li Y., Review perspectives for natural working fluids in china, International journal of refrigeration, Vol. 30, 2007, 568-581.

[6] Fan Y., Luo L., Souyri B., Review of solar absorption refrigeration technology: Development and applications. Renewable and sustainable Energy Reviews, vol.11, 2007, 1758-1775.

[7] Sun J. , Fu L., Zhang S., A review of working fluids of absorption cycle. Renewable & Sustainable Energy Reviews, Vol. 16 , 2012, 1899-1906.

[8] Alexis GODEFROY, Nathalie MAZET, PhD thesis, "Analyse thermodynamique et performances dynamiques de cycles hybrides impliquant des procèdes a sorption", 2020,

page 38-40

[9] Koyfman U. ,Jelinek M. ,Levy U.,Borde J., An experimental investigation of absorptiondiffusion cycle working with organic fluids, Applied Thermal Energy, Vol.23, 2003,1881-1894

[10]Karamangil M. ,Coskun S. ,Kaynakli O. Yamankaradeniz N. A simulation study of performance evaluation of single-stage absorption refrigeration system using conventional working fluids and alternatives. Renewable & Sustainable Energy Reviews, Vol. 14, 2010, 1969-1978.

[11]Rodríguez-Muñoz JL. , Belman-Flores JM., Review of absorption diffusion refrigeration

technology, Renewable & Sustainable Energy Reviews, Vol. 30 , 2014, 145153

[12] Zhen L. ,Yong L. . ,Huashan Li ,Xianbiao Bu ., Weibin M. , study of performance of absorption-diffusion cycle working with TFE-TEGDME mixture, Building Energy, Vol.58 , 2013, 86-92

[13] Makhloul M., Kherris S., Chadouli R. and Asnoun A. Amélioration de la performance d'un cycle frigorifique à absorption-diffusion NH3-H2O-H2, Revue des Energies Renouvelables Vol. 12 N°2 (2009) 215 - 224

[14] Kherris M., Makhlouf S., Chadouli R., Asnoun A. Performance improvement of a NH3-H2O-H2 absorption-diffusion refrigeration cycle. Revue des Energies Renouvelables 2009;12:215-24.

[15] Batoul B., kaabi A.N, khattib Y., Belhamri A., Gomri R. Thesis: Thermodynamic study of refrigerant mixtures and their use in trithermal machines, 2012.

[16] Chekir N., Bellagi A. Separation in the generator of an absorption refrigeration machine: thermodynamic modelling, calculation of liquid-vapour equilibria and study of the case of binary mixtures of n-alkanes ,2000, 10-12

[17] Chekir N., Mejbri Kh. , Bellagi A., Simulation d'un refroidisseur à absorption fonctionnant avec des mélanges d'alcanes, Revue Internationale du Froid, Vol.29, 2006, 469-475

[18] Dardour H., Mhiri H., Gabsi S., Bellagi A, Etude des machines frigorifiques a absorption diffusion utilisant un mélange d'alcanes : étude systématique et modélisation rigoureuse de l'absorbeur ,2012 .

[19] Website of refrigeration specialist ABC CLIM Autoentrepreneur, url: https://www.abcclim.net/les- gwp-des-fluides.html, July, 2017.

[20] Chaouachi B., PhD thesis, "Contribution à l'étude des cycles à absorption. Performances du mélange Chlorure de lithium - chlorure de magnésium, d'un générateur solaire et d'un échangeur multitubulaire" 1996, pages 13-14.

[21] James M. Calm ,The next generation of refrigerants - Historical review, considerations and outlook ,2008.

[22] Bolaji B., The choice of environmentally friendly refrigerants and current alternatives in vapour compression refrigeration systems. Journal of Science and Management, vol.1, 2011, 22-26.

[23] Sun J., Fu L. Zhang S. A review of working fluids of absorption cycle. Renewable & Sustainable Energy Reviews, Vol. 16 , 2012, 1899-1906

[24] Von Platen. Absorption-diffusion cycle patent.

[25] Devecioglua A. G. , Oruça V. Characteristics of Some New Generation Refrigerants with Low GWP. Energy Procedia 75 (2015) 1452 - 1457.

[26] K. Harby, Hydrocarbons and their mixtures as alternatives to environmentally unfriendlyhalogenatedrefrigerants : Anupdatedoverview Renewable and Sustainable Energy Reviews 73 (2017) 1247-1264.

[27] Ben Ezzine N., Garma R. ,Bourouis M. , Bellagi A., Experimental studies on bubble pump operated light hydrocarbon diffusion absorption machine for solar cooling, renewable Energy, Vol. 35 , 2010, 464-470

Printed by Books on Demand GmbH, Norderstedt / Germany